Environmental Awareness:
ACID RAIN

AUTHOR
By Mary Ellen Snodgrass

EDITED BY
Jody James, Editorial Consultant
Janet Wolanin, Environmental Consultant

DESIGNED AND ILLUSTRATED BY
Vista III Design, Inc.

 BANCROFT-SAGE PUBLISHING, INC.
601 Elkcam Circle, Suite C-7, P.O. Box 355
Marco, Florida 33969-0355

Library of Congress Cataloging-in-Publication Data

Snodgrass, Mary Ellen.
    Environmental awareness—acid rain / by Mary Ellen Snodgrass;
edited by Jody James, Editorial Consultant; Janet Wolanin,
Environmental Science Consultant; illustrated by Vista III Design.
    p. cm.—(Environmental awareness)
    Includes index.
    Summary: Examines acid rain, its effects on the world, and possible
ways of stopping such damage.
    ISBN 0-944280-30-7
    1. Acid rain—Environmental aspects—Juvenile literature. 2. Air-
Pollution—Juvenile literature. [1. Acid rain. 2. Air-Pollution. 3. Pollution.]
I. James, Jody, Wolanin, Janet. II. Vista III Design. III. Title. IV. Title: Acid
rain. V. Series: Snodgrass, Mary Ellen. Environmental awareness.
TD195.44.S65   1991
363.73'86—dc20

International Standard Book Number:
Library Binding 0-944280-30-7

Library of Congress Catalog Card Number:
90-26255
CIP
AC

PHOTO CREDITS

COVER: Vista III Design; Nancy Ferguson p. 10, 26, 40; NASA p. 30; Unicorn
Photography, Deneve F. Bunde p. 7, Robert Hitchman p. 32, Wayne Floyd p. 37;
Vista III Design, Ginger Gilderhus p. 14, 17, 34, Grant Gilderhus p. 4, 8, 19, 21, 23,
24, 25, 39, 42, Jackie Larson p. 13.

TABLE OF CONTENTS

Pure rainwater is necessary to support life on Earth.

4

CHAPTER ONE

A THREAT FROM THE SKY

Most people look forward to an occasional rainy day. Rain is a source of water for lawns, trees, and flowers. It wets the farmer's crops and helps them grow. It makes puddles and streams in which frogs, toads, and lizards can dampen their skins. In cities, rain cleans the streets and settles the dust. After a long, hot summer day, a pleasant shower or thunderstorm cools the air.

Not all rain is pleasant, however. Since the 1870s, scientists have known that some rain is harmful. Rain that carries dangerous chemicals can harm plants, animals, and people. It can hurt rivers, lakes, and soil. In cities, this harmful rain damages buildings, cars, and statues. This rain is called **acid rain**.

Here is how three students at Bentham Elementary School learned about acid rain.

A SPECIAL ASSIGNMENT

The dismissal bell rang, sending students hurrying home for the day. Lockers slammed. Voices called good-byes. Soccer players and cheerleaders put on their uniforms and began practicing.

Sylvia, Cam, and Pete met in Mr. Cantrell's room to discuss their science project. As project leader, Sylvia assigned tasks to the other team members.

"Cam, would you like to be the one to gather leaves and pine cones? You live near the park," said Sylvia. She wrote Cam's initials beside Part One of the assignment.

"That sounds okay to me," said Cam with a nod. "My little brother will have a good time helping me find different kinds of trees."

"What can I do to help?" Pete asked.

"Let's see, Pete," Sylvia said as she read over the list of responsibilities. "This assignment calls for posters and headings. Can you make neat letters?"

"Sure. I'm good at that," replied Pete. "I did all the charts for the last fall festival."

"Then if you make the posters and Cam gathers the samples, writing the results are left for me to do," concluded Sylvia. "I think the three of us will finish this project on time and earn a good grade."

"Do you mind writing the descriptions?" Cam asked.

"No. That's just fine with me. I like to put ideas together," Sylvia said, placing her pen in the end of her spiral notebook. "If the work gets to be too much for me, you and Pete can help out," she added.

"Sure," said Pete. "We'll both help with the writing."

The group turned to Mr. Cantrell's back tables. They began looking over projects that last year's classes completed.

"Well, I'm glad you're all here," Mr. Cantrell said with a smile as he came from the supply room. He began pulling folders out of his leather briefcase and selected one marked "Acid Rain." He laid the folder on the table in front of the three students. "I have an unusual assignment for this group."

HARMFUL RAIN

"I thought we were going to do a study of the trees that are common to Westfield County," Pete began. "This folder says 'Acid Rain.' Have you changed your mind about what you want us to do?"

"No, Pete," Mr. Cantrell replied, taking a seat at the end of the work table. "I recently attended a conference at the Southeastern Environmental Center. I learned some disturbing facts there about the trees in this part of the county. Many of them are dying. I hope that your group can find out more about the problem and educate the whole class—me included."

"What is causing them to die, Mr. Cantrell?" Sylvia asked. "Do they have a disease?"

"Not exactly, Sylvia," Mr. Cantrell replied. "Scientists have been testing the rain and snow in this part of Colorado. Studies show that perhaps smoke from chimneys as far away as the Ohio Valley could mix with the rain and snow. When the moisture falls, it is much too acidic for plants."

Acid rain and snow is causing trees to die in many parts of the United States and Canada.

Acid rain soaks into plant and tree leaves removing the protective surface. Often the rain will scar and weaken the leaves and roots causing the plants to die.

"Why is acid rain harmful?" Pete asked with a puzzled expression on his face. "I thought that rain was good for plants."

"Well, normal rain is essential for plants, Pete, but this is not normal rain," Mr. Cantrell replied. "When it soaks into leaves, it removes the protective surface. Often the acid rain scars and weakens the leaves and roots of plants. I want your group to examine as many types of leaves as you can. I want you to look for signs of acid rain damage. Find out what causes acid rain and what we can do to prevent it."

"Good grief," moaned Cam. "There's air **pollution** to worry about, and too much garbage. Now we've got to worry about the rain. How can three students keep acid rain from harming trees?"

The roots on this plant have been damaged by acid rain causing the leaves to wilt.

*Acid rain caused by air pollution falls into
rivers, lakes and streams destroying plant and fish life.*

CHAPTER TWO

ACID RAIN AND NATURE

Do you have an answer for Cam? Did you know that some types of air pollution mix with rain? Did you ever wonder where air pollution goes after it falls to the earth? Many people are starting to ask questions about pollutants in rainwater. They are worried because acid rain can kill plants, fish, and birds.

Without plants and animals, the earth would be empty. If acid rain continues to destroy nature, humans will soon be in trouble. In order to understand acid rain, we must first learn a little about acids and bases.

ACIDS AND BASES

Acids are common chemicals. They taste sour and come in varying strengths. (WARNING: *Do not taste any unknown substance to see if it is an acid. Many acids are strong enough to cause severe burns.)* Many

acids are formed in nature. Oak leaves, for example, contain a great amount of acid. Early settlers in America used these leaves to tan and soften leather.

Oak leaves are an example of acid formed in nature.

11

Acids can also be useful in making fertilizer and in the preparation of gasoline and kerosene from raw petroleum. Other acids are used to make dyes, paper, and explosives. Some acids are strong enough to dissolve metals such as copper and chromium.

Many acids are used in business and industry. Nitric acid is an industrial-strength cleaning agent. It is used to scour tanker trucks and make them clean enough to carry milk and other foods. Weak acids are often used in medicine. Boric acid, for example, is used to rinse eyes to remove irritants and to soothe itching and stinging.

Most people use acids in the kitchen and laundry. Cooks use acids like lemon juice and vinegar for many purposes. Vinegar tenderizes meat fibers and makes them easier to chew and digest. Both vinegar and lemon juice bring out the taste of vegetables, seafood, and salad greens. Also, vinegar helps preserve pickles so they stay crunchy and flavorful. Chlorine bleach is an acid that helps whiten and deodorize laundry. It also kills germs, mildew, and mold on bathroom fixtures and sinks.

Another acid, **gastric acid**, is manufactured in the stomach. This acid helps break food down so that the body can be properly nourished. Without stomach acids, food could not nourish the body. To protect delicate body parts from the acid, the body also produces mucus. The mucus covers the organs so that gastric acid does not harm the tissues.

Bases are another common group of chemicals. Bases feel slippery or soapy on the hands. Limestone is a base found commonly in nature. To grow better crops, farmers may sprinkle powdered or granular limestone on their fields. This limestone releases other minerals that make plants healthy.

Stronger bases are used to make soap and other cleaning solutions found in the home. Some common weaker bases that people use at home include baking soda, ammonia, and remedies for indigestion.

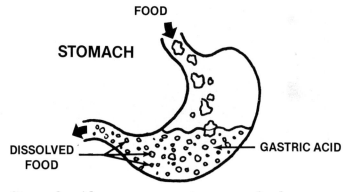

Stomach acids are necessary to process food and nourish the body.

Farmers sprinkle limestone on their fields to help produce healthy plants and better crops.

Litmus paper tests help people determine the nature of an unknown substance. Litmus paper turns red when dipped into an acid and blue when dipped in a base substance.

TESTING FOR ACIDS AND BASES

Litmus paper helps people determine the nature of an unknown substance. Litmus is a natural testing agent that comes from a type of moss that grows on rocks and trees. It shows whether a substance is acidic or basic. When litmus paper is dipped into an acid, the paper turns red. When litmus paper is dipped into a basic solution, the paper turns blue.

When scientists test substances for the presence of acids or bases, they chart the results on a line of numbers. This line is called a **pH scale**. The numbers range from 0 to 14. Substances that fall at the low end of the scale are acidic. Those that fall at the high end are basic. Substances that fall close to the middle, are **neutral**. They are neither acidic nor basic. Distilled water is the most common neutral substance.

When acids and bases are mixed, they counteract each other. In other words, the strong qualities of the acid cancel the strong qualities of the base. When this cancellation occurs, they are said to **neutralize** each other. A common example of neutralization is the use of milk of magnesia or antacid mints to relieve an upset stomach. When the medicine, which is a base, mixes with an abundance of stomach acid, the two substances cancel each other. The acid no longer causes stomach discomfort.

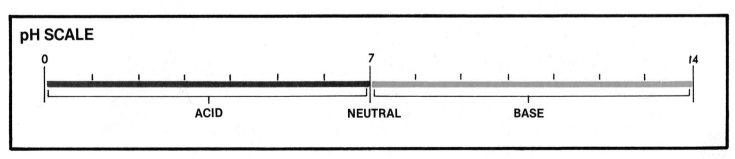

pH SCALE

0   7   14

ACID   NEUTRAL   BASE

Scientists use a pH scale to test substances for the presence of acids and bases.

ACID IN RAIN

Some pollutants in the air can mix with rainwater to form an acid. These pollutants are frequently composed of nitrogen or sulfur. When water mixes with **nitrogen oxide**, it produces **nitric acid**. When water mixes with **sulfur dioxide**, it forms **sulfuric acid**. Both nitric acid and sulfuric acid have very low pH values. These low values mean that the two acids are strong.

When cities are filled with cars and trucks, exhaust fumes dirty the air. When water in the air mixes with the fumes and falls to the ground, the result is acid **precipitation**, commonly called acid rain. Cars are not the only creators of dirty air. Factory smokestacks also carry large amounts of dirty particles. A common type of dirty particle is sulfur from burning coal. The coal smoke is often black and has a strong odor. When coal smoke is mixed with precipitation, it forms a very strong kind of acid rain.

Coal smoke when mixed with precipitation forms a strong kind of acid rain.

Exhaust fumes from cars and trucks are one cause of acid rain.

THE EFFECTS OF ACID RAIN

Acid rain affects every living thing on the earth. It harms animals, plants, and humans. It even harms the surface of the earth.

ACID RAIN AND WILDLIFE

When the pH of ponds and streams begins to fall toward the acidic side of the pH scale, fish, frogs, toads, and insects are affected. Their protective shells or skin may weaken. Some animals are unable to lay eggs. Others produce healthy eggs, but acid rain **stunts** the newborn creatures. In some fish, acid water weakens the skeleton or causes abnormal growths on the bones. Many animals and insects die from these changes in their bodies. Others stop reproducing. Soon, the small animals and insects that provide food for larger creatures are in short supply.

On the surface, **wetlands** may not seem to be harmed by acidic water. Ponds and lakes may mirror a crystal blue. Beneath the surface, however, is another story. As the water becomes more acidic, green **algae** may collect on the bottom. These simple green plants produce a filmy undergrowth of interwoven branches.

Algae are useful because they provide food for fish. Humans eat some types of algae. They use others to thicken ice cream, pudding, cheese, jam, and other foods. Processed algae makes a smooth base for cosmetics, leather finishes, car polish, paint, and **insecticides.**

Why, then, are algae harmful? Like all life forms, algae must exist in a balance with plants and animals. If acidic water causes an overgrowth of algae, other life forms may be smothered at the bottom of lakes and ponds. Rotting algae releases a poisonous gas that destroys shellfish. Some of the gas bubbles to the surface, creating foul odors. The body of water becomes too choked with algae to support life and soon it dies.

ACID RAIN AND PLANT LIFE

After normal rain rinses the air free of pollution, it may continue as clean precipitation. It helps **dilute** the acid rain that has already fallen. By diluting the acid, the rain makes the acid weaker. Clean rain also rinses leaves and moves the acidic deposits into the ground.

When a body of water becomes too choked
with green algae, plant and animal life soon dies.

Acid rain in the soil soaks the roots of trees and shrubs. The acidic water is drawn up into the body of the plants. Too much acid is not good for plants.

If acid rain is not rinsed quickly enough from leaves and needles, it can eat through the waxy coating on their surface. Without this protective coating, leaves may be scorched by the sun. Evergreens, such as pines and spruce trees, may drop their needles. The trees become weak and sick. Diseases, insects, and **fungi** can attack weakened trees and kill them.

Plants that have no leaves stop producing oxygen. Since we need the oxygen that the plants produce to breathe, we, too, suffer from acid rain. Soon the fresh air that healthy forests create is less fresh. The loss of air quality makes the world a little less pleasant and a little less friendly to human life.

Often acid rain does not soak into the inner leaves or needles of trees. Then only the tops and outer edges are affected. The growth of these trees may be stunted. They may not produce new shoots. Their color and shape may become abnormal. Such trees are not appealing to visitors of parks, campsites, and playgrounds.

Acid rain also causes other problems. When leaves fall in acidic water, they are pickled, like cucumbers in a jar of vinegar. Instead of breaking up into **humus** and fertilizing the earth, they choke waterways and streambeds. Acid rain also causes limbs and twigs to die and fall to the ground. This often results in a fire hazard. The fires, once started, can destroy whole forests.

ACID RAIN AND THE ENVIRONMENT

When forests are weakened by acid rain, the **environment** is affected. Disastrous forest fires leave the land unprotected. No leaves or limbs remain to soften the fall of heavy rains. No roots are present to hold ground moisture. The end result is rapid **runoff** of rainwater and **soil erosion**.

Much of the good soil washes away into creeks and rivers. Open gullies of **hardpan** or even **bedrock** are all that remain. The soft accumulation of soil and humus that usually covers the hardpan and bedrock is lost. The layer of soil that filters out **ground water** pollution is lost. Soon, the water that fills wells and **reservoirs** may be unfit for drinking or swimming.

Eroded land offers no secure nesting places for small animals.

Eroded land and burned-out forests offer no secure nesting places for small animals, such as chipmunks and squirrels. As the animals and insect populations die off, much of nature's beauty is destroyed. A land in which animals and insects cannot live may soon become unfit for humans as well.

ACID RAIN AND HUMAN LIFE

Another dangerous result of acid rain is its direct effect on the human body. Scientists have evidence that connects acid rain with severe diseases of the lungs. Acid rain may cause bronchitis, asthma, and lung cancer when people inhale the acidic moisture in the air. Acid rain especially affects children, old people, and people who have heart disease. It may even affect infants before birth.

Acid rain also causes other human health problems. These occur because of the **leaching** of dangerous substances from the soil. When strongly acidic water soaks into the earth, it dissolves ground metals. Some of these, known as **heavy metals**, may be hazardous to animals and humans.

Heavy metals may enter the body through polluted foods and water. Metals such as zinc, mercury, arsenic, and lead collect in the soft tissues of the human body, particularly in the kidneys and liver. They damage nerve cells and destroy brain tissue.

Acid rain can wash these dangerous metals from the ground into ponds and lakes. When people eat fish, shellfish, or game that contains heavy metals, they may suffer metal poisoning. Unfortunately, their symptoms may not be clear to health workers. The disease may reach a life-threatening stage before it is finally diagnosed. Some people may even die before they can be helped.

Acid rain can also affect city water systems and wells. As the low pH liquid eats into drinking water pipes, the acid may dissolve their metals. Soon, dissolved lead, aluminum, mercury, and copper from the pipes may be consumed by people who are unaware of the danger.

ACID RAIN ON CITIES AND FARMS

The dissolving action of acid rain is particularly harmful in cities. When low pH water pelts buildings and monuments, it breaks down marble, brick, wood, and cloth fibers.

The photos above show how acid rain can destroy metal, wood and brick.

*Acid rain is a threat to monuments such
as the Lincoln Memorial in the United States.*

Historic statues and decorations on buildings begin to look like sponges because of the small holes in them. Without restoration, these monuments will lose their original shape. If acid rain continues to soak cities, the world will lose many famous monuments, such as the Statue of Liberty and the Lincoln Memorial in the United States.

Acid rain also affects the paint, glass, leather, and rubber in cars, trucks, and buses. It may cause pits to form in paint and broken fibers to occur in fabric. It may eat into tires and weaken them. The cost of replacing everything that acid rain ruins is high.

Farms, too, suffer when acid rain falls regularly. Vegetables such as cabbage, beans, tomatoes, and broccoli become weak. They suffer the same dangers that affect forest leaves. The waxy surface that protects delicate vegetable tissue is destroyed. Because of this weakening, disease, fungi, and insects can attack the inner tissues. Leaves develop ugly growths and crops stop growing. Because of the effects of acid rain on fields of vegetables, farmers lose money. Because the farmers have fewer vegetables to sell, they have to raise their prices. In some areas, food shortages may occur.

The quality of fruits and vegetables that people eat may suffer from acid rain.

Acid rain is formed when polluted air from factory smokestacks mixes with rain or snow.

CHAPTER THREE

LEARNING ABOUT ACID RAIN

Before scientists could begin testing for acid in rain, they had to be aware that acid rain existed. They had to connect the low pH in precipitation with pollution in the air. After this information became widely known, people began to pressure the government agencies to solve the problem of acid rain.

THE DISCOVERY OF ACID RAIN

In the seventeenth century, scientists first realized that all kinds of precipitation mixed with particles in the air. In 1872, Robert Angus Smith, an English chemist, invented the term "acid rain" to describe the mixture. Much later, in the late 1960s, a soil scientist named Svante Oden wrote about acid rain in Swedish newspapers. Other scientists became interested in his work.

The scientific community of northern Europe did the most important work. Concerned individuals from Sweden, Finland, Denmark, and Norway made the connection between acid lakes, fish kills, and the many factories of central Europe. They realized that wind carries smoke and polluted air from factory chimneys. The scientists analyzed the water in wetlands. They discovered that the smoke was dissolving in rainwater. When the water fell to the earth far to the north, it was too acidic. The acid rain was destroying lakes and forests.

Since that time, Canadian scientists have found that North America is suffering a similar problem with acid rain. Polluted air from smokestacks in the United States drifts north into Canadian air space. Rain, snow, and other forms of precipitation mix with the particles and create acid rain. In this way, the smoke from a steel mill in Cincinnati, Ohio, may be stunting the growth of spruce trees in Saskatchewan. The car exhaust from a New Jersey highway may be killing the fish in Ontario. In addition to harming the environment, acid rain strains relationships between the United States and Canada. If the United States wants to remain a good neighbor to Canada, it must clean up its polluted air.

TAKING SAMPLES

To measure and control the pollution that results in acid rain, scientists must take many samples. They must test the makeup of soil, air, and water as well as of plant life. To test for air pollution, workers in many cities place air sampling machines throughout the area. Some machines are close to the ground. Others are high up on buildings and towers. Each of these machines contains a paper **filter**. When air is drawn through the filter, the machine draws a graph of the amount of solid particles in the air.

The location of each machine helps workers determine where pollution is most common. By looking at the graphs, workers can see when the largest amount of solid particles occurs. The graphs also show any increase in the number of particles during a particular type of wind or weather condition, such as right before a storm or on foggy autumn days. This information helps city and industrial planners understand how and when to reduce the most severe types of air pollution that cause acid rain.

Workers can also learn about kinds of pollutants by using air sampling machines. They collect the paper filters and test the particles that are trapped in the paper. By using computers that identify the chemicals in each sample, the workers find out what kinds of particles are in the air. Some laboratory testing may point to dangerous heavy metals. Other testing may locate nitrogen and sulfur. All of these substances can create acid rain.

To test for increases in acid rain, scientists often take **core samples**. By forcing a long, hollow tube into the bottom of a lake or stream, the scientists can draw out a core of mud. They can study the core to tell how acidic the rain was long ago. They can figure out where acid rain is increasing and how fast. By comparing bits of decayed matter from plants and animals, they may also learn what life forms once lived in the area. If many of the plants and animals are no longer living there, one reason may be the lowered pH from acid rain.

There are other methods of locating acid rain damage. One method is to sample grass, leaves, roots, and stems, as well as the bones, scales, feathers, and other tissues of animals. By learning how much heavy metal or other chemicals the samples contain, laboratory workers can decide how much danger there is to human life. If plants, fish, birds, and animals are seriously contaminated, agents of the **Environmental Protection Agency [EPA]** may post warning signs. The signs tell visitors that berries, mushrooms, eggs, oysters, mussels, clams, and meat taken from the area are not fit to eat.

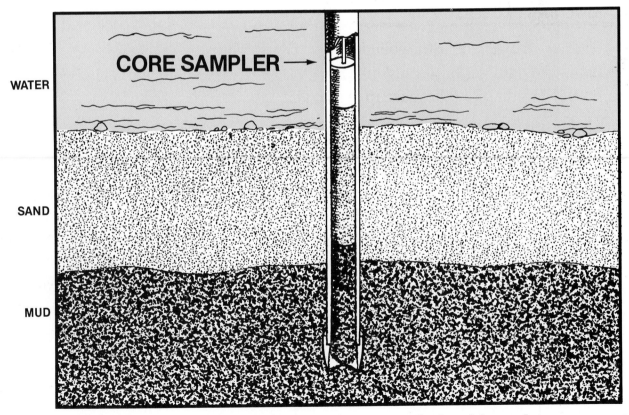

To test for increases in acid rain, scientists take core samples from lakes and streams.

Everyone on Earth must work together to find solutions to the effects of acid rain.

CHAPTER FOUR

TAKING ACTION

Locating the sources of the pollution that causes acid rain is only the beginning. To stop acid rain and heal the earth, everyone —scientists, factory owners, politicians, and individuals—need to work together on solutions. No one can afford to ignore the problem or to do nothing to solve it. There is only one Earth and we all live on it together.

ROADBLOCKS TO CONTROLLING ACID RAIN

Acid rain is difficult to control. There are two main reasons for this. One reason is that no one can say for sure where a given batch of acid rain came from. The second reason is that many people do not believe that acid rain is an important problem.

FINDING SOURCES OF ACID RAIN

Because wind travels in many directions, it is hard for pollution samplers to discover the exact source of air pollution. When acid rain stimulates the growth of algae in the Great Lakes, no one can say for sure that it came from high sulphur coal being burned in Kentucky. When acid rain cripples a pine forest in Maine, no test can point to a particular furniture factory in Virginia.

Because tracing the exact causes of air pollution is impossible, government agencies often hesitate to require changes that might improve the quality of air. The agencies think it is unfair to blame factories without clear evidence of wrongdoing.

Disagreements between anti-pollution supporters and government agencies often result in inaction. Unfortunately, while no one

Some types of pollution from nature such as smoke from this volcanoe cannot be controlled.

is stopping the problem, air pollution continues to dirty the air. And dirty air continues to mix with rain which falls to the earth.

While people argue fine points concerning irregular acid levels, many of them are missing the main issue: acid rain is deadly. Whatever is killing fish and other wildlife is dangerous to humans as well. To keep the earth clean enough for life to go on, we must create ways to stop acid rain.

CLEANING THE AIR

One of the most obvious facts about acid rain is that it is created in the air. Once acid rain reaches the earth, there is little that people can do to halt its effects on living things. To improve the pH level of precipitation and control acid rain, people must reduce air pollution. No one can control volcano smoke but we can control our car and factory smoke.

CONVINCING PEOPLE OF THE DANGER

Convincing people that air pollution causes acid rain is a big problem. Some say that rain has always contained acid, even before factories were built. It is true that volcanoes, geysers, decaying plants, forest fires, and smoke from fireplaces and barbecue grills add acid to the air. But factory smoke and car exhaust create a much more serious threat than these natural forms of pollution. The increase in acid rain in recent years reflects the increase in factories and cars more than any other factor.

Other people get caught up in questions about the fine points of acid rain. One such point is the irregularity of high acid levels. Sometimes acid levels in wetlands reach a dangerously low pH level. These changes in acidity cause large fish kills. One explanation of these disasters is the coming of spring. As the air warms, snow begins to melt. If the snow contains a large amount of acid, the acid enters ponds and streams all at one time. Immediately, the pH of the water drops. The acid makes the environment too unhealthy for fish to live. The dramatic drop in pH causes the fish to die all at once.

Catalytic converters are required on all American-made vehicles. These converters change dangerous exhaust particles into safe exhaust.

SOLUTIONS WITHIN THE LOCAL COMMUNITY

One of the dirtiest and most common polluters is the car engine. Engines that burn **fossil fuels** such as gasoline, kerosene, and diesel fuel must be kept in good running condition. By 1975, most American-made cars had **catalytic converters** in them. These converters change dangerous exhaust particles into safe exhaust. In the early 1980s, the government made it illegal to remove a catalytic converter from a car. This regulation has helped reduce the amount of nitric acid and sulfuric acid in the air, but much still remains to be done.

One method of reducing automobile exhaust is to change to engines that burn alcohol made from corn. Alcohol and other alternative fuels do not produce the particles that form acid rain.

Another way to improve air quality near highways is to reduce traffic snarls. If cars and trucks move at a reasonable speed, they do not stand still and waste fuel. A better method of speeding people to jobs and classrooms is to encourage public transportation. People who ride buses, trains, and subways reduce the number of cars on the road. In places where public transportation is limited, people can reduce traffic by carpooling. Some cities provide safe and inexpensive park-and-ride locations where people can park their cars and take a public van or bus to work.

Everyone can help clean the air and reduce acid rain. Instead of burning leaves and tree limbs, people can **compost** them. In addition, by **recycling** or taking solid waste to a **landfill** instead of burning it, people can further reduce particles in the air.

Finally, people should check **flues** and chimneys on fireplaces and wood stoves. When soot builds up inside the flues, it produces a thick smoke that contains harmful acids. People should never burn treated timber, colored or metallic paper, foam products, dangerous chemicals, or plastics. By burning only clean, dry logs and kindling, people can help control acid rain.

35

Factories that have **recovery systems** actually vacuum the gas as it escapes. These systems cool hot gas, remove valuable chemicals, and collect them in barrels. The chemicals can then be reused. Such systems reduce the amount of acid-causing gas that escapes to mix with rain and form acid rain.

Controlling fuels and raw materials can also reduce factory exhaust. Some coal-burning factories now use coal that has been washed and crushed or that has a low level of sulfur. This exhaust reduction cuts down on the formation of sulfuric acid in precipitation. Other companies are changing to less harmful solids, liquids, and gases in the manufacturing process.

Using alternate forms of energy may be a good way of reducing acid rain from air pollution. Alternate forms of energy include **nuclear energy** and **solar energy** as well as energy from the wind and moving water. However, none of these alternatives are yet able to replace fossil fuels entirely. Also, nuclear power may be even more hazardous than acid rain.

Factory and power plant smokestacks belch out great quantities of black smoke, soot, dust, and other unhealthy particles. For awhile, anti-pollution groups recommended higher smokestacks to spread the pollutants over a wider area. This method keeps polluted air from poisoning the immediate area, but it does not solve the problem. A better idea than tall smokestacks is the **scrubber**. This device contains a lining made of **activated charcoal**. The scrubber traps harmful particles and changes them into safer **by-products**.

Many other methods are available to control factory-caused air pollution. A simple device that helps keep particles out of the air is a cap to trap **fly ash**. A more up-to-date device is the **flare tip**, which injects steam into the gas as it leaves the smokestack. The addition of steam helps reduce the amount of smoke. Sometimes the steam eliminates pollutants entirely. Another modern device magnetizes particles. Like magnets, the particles stick together and fall harmlessly out of the way of escaping smoke.

Solar energy panels are used as an alternate form of energy to help heat homes, businesses and factories.

Scientists have not yet invented systems that make solar, wind, and water energy practical on a large scale. Until scientists know more about these sources of energy, they may have to continue making coal and oil furnaces safer. Whatever methods they choose, factory owners who make efforts to reduce acid rain show their concern for the environment.

REVERSING THE DAMAGE

Where acid rain is already a serious problem, people can take measures to raise the pH level of the water and soil. One method is to sprinkle limestone over the area. This can be done at ground level or by boat or airplane. Because limestone is a base, it neutralizes the acids in the water.

Unfortunately, applying limestone is not a cure-all for the problem of acid rain. Because large amounts of bases can also harm the environment, limestone must be introduced carefully into water and soil. Also, scientists cannot create sudden, violent changes in the pH level. They must give fish, birds, and plants a chance to adapt to a rise in the pH level. Scientists must be very careful not to do more harm than good when trying to neutralize acid rain.

Another method that may help mildly acidic areas is to flush them with clean water or water mixed with a base. By rinsing away the acids, diluting them, or neutralizing them with a base, a cleanup operation could rescue a small area saturated with polluted rain. This method might be applied to a pond where melting acid snow has lowered the pH. Because this method requires much preparation and cost, it is not suitable for widespread use.

*When scientists attempt to neutralize the environment by
adding limestone, plants, fish and birds must be given a chance
to slowly adapt to a rise in the pH level.*

39

***To help protect the environment you might
ride a bicycle to school or work.***

CHAPTER FIVE

THE INDIVIDUAL'S PART

The three students in Mr. Cantrell's class did a thorough job of researching acid rain. They collected samples of leaves that had been damaged by acid rain in their area. They also talked to local authorities and environmental groups. At Sylvia's suggestion, they concentrated on finding out what an individual can do to help stop acid rain.

Scientists told the students that no one person alone can halt acid rain. Solving the acid rain problem requires the cooperation of factories, government authorities, and entire communities. Still, each person can help. Here are the lists of suggestions the students prepared for their class.

PROTECT THE ENVIRONMENT

1. Take paper, cardboard, glass, aluminum, and plastic to recycling centers to be reused rather than burned.

2. Dispose of paint, used oil, and other pollutants according to local laws.

3. Compost dead leaves and limbs instead of burning them.

4. Use fireplaces and grills properly. Never burn trash in them.

5. Keep chimneys clean.

6. Share rides or use public transportation to reduce the use of cars.

7. Use alternate sources of energy, such as wind or solar energy.

8. Walk or ride your bicycle.

9. Ask questions. Read articles about pollution and its effect on acid rain. By learning more, you can be an effective voice to protect the environment.

While out enjoying nature, watch streams, lakes, rivers and ponds for any unusual changes such as fish kills or sour smells.

ENCOURAGE GOVERNMENT OFFICIALS TO TAKE ACTION

1. Write or call local officials or state or national representatives. Encourage them to vote for stronger anti-pollution laws. Insist that they enforce the laws we already have.
2. Join or support citizen's groups that guard against acid rain. Insist on heavy fines for polluters.
3. Observe streams, lakes, rivers, and ponds near your home. Report unusual changes, such as fish kills or foul smells.
4. Support widespread testing for the effects of acid rain.

ENCOURAGE INDUSTRY TO KEEP AIR FREE OF POLLUTION

1. Avoid products made by manufacturers who pollute or refuse to cooperate in the cleanup and protection of the air.
2. Report dangerous or suspicious smoke and fumes to EPA officials.

GLOSSARY

acids (A sihds)   sour substances that have low pH values, can neutralize bases, and can dissolve metals

acid rain (A sihd RAYN)   rain that combines with acids in air pollution and is therefore more acidic than normal rain

activated charcoal (AK tih vay tihd CHAHR kohl)   charcoal that has many pores, or openings, in which to trap particles

algae (AL Jee)   small green plants that grow in water and sometimes form mats on lakes or ponds

bases (BAYS ehs)   substances that feel slippery or soapy, that can neutralize acids, and have high pH values

bedrock (BEHD rahk)   solid rock that lies below soil

by-products (BY prahd uhkts)   waste products made by a factory while it is creating useful goods

catalytic converters (kat uh LIHT ihk kuhn VUHRT uhrz) devices that clean the exhaust from cars, trucks, and other vehicles

compost (KAHM pohst)   a pile of leaves, grass clippings, and green waste that will break down into useful mulch

core samples (KOHR SAM puhlz)   cutout samples of soil or mud that show the make-up of an area such as a lake bottom

dilute (dih LOOT)   to add liquid to a strong substance to make it weaker

environment (ihn VYRN mihnt)   the world in which living things live and grow

Environmental Protection Agency (EPA)   a federal agency that tries to help protect the environment in the United States

filter (FIHL tuhr)   a device like a strainer that separates solid particles from a gas or liquid

flare tip (FLAYR TIHP)   a device in a smokestack that injects steam into fumes to lower the number of escaping particles

flues (FLOOS)   a tube in a chimney, through which smoke is drawn off

fly ash (FLY ASH)   solid particles of dust, soot, and ash which are blown from a fire

fossil fuels (FAHS ihl FYO elz)   fuels such as coal, oil, and natural gas that have formed over millions of years from decayed plants and animals

fungi (FUHN jy)   small organisms such as mildew, yeast, mushrooms, and mold

gastric acid (GAS trihk A sihd)   acid that forms naturally in the stomach and helps digest food

ground water (GROWND waht uhr)   water that collects below the earth's surface

hardpan (HAHRD pan)   soil that is pressed so tightly together that roots cannot grow in it

heavy metals (HEHV ee MEHT uhlz)   metals, including iron, zinc, mercury, lead, chromium, and arsenic, some of which can poison humans and animals

humus (HYOO muhs)   any substance that was alive at one time, such as decayed leaves or animal bones and teeth

insecticides (ihn SEHK tih syd)   a chemical that kills insects

landfill (LAND fihl)   a place where solid waste is buried in a safe manner between layers of dirt or clay

leaching (LEECH ihng)   dissolving metals from the soil into water supplies or ground water

litmus paper (LIHT muhs PAY puhr)   paper made from a moss that turns pink in acidic solutions and blue in basic solutions

neutral (NOO truhl)   neither acidic nor basic

neutralize (NOO truh lyz)   to mix an acid with a base so that the strong qualities of one cancel the strong qualities of the other

nitric acid (NY trihk A sihd)   a strong acid formed when nitrogen oxide from car exhaust mixes with water

nitrogen oxide (NY troh jihn AHK syd)   a gas found in car exhaust

nuclear energy (NOO klee uhr IHN uhr jee)   energy that is released when atoms are split or forced to combine

pH scale (PEE AYCH SKAYL)   a method of comparing substances that are acidic and basic

pollution (puh LOO shuhn)   dirtying of the environment

precipitation (pree sihp uh TAY shuhn)   any form of water that falls from the skies, including rain, hail, sleet, snow, and fog

recovery systems (ree KUHV ree SIHS tuhmz)   systems used by factories to collect reusable chemicals from smoke before the smoke leaves the factory

recycling (ree SY klihng)   reusing materials, such as newspaper, glass, and aluminum cans

reservoir (REH suhrv oyr)   an artificial lake where water is collected for later use

runoff (RUHN ahf)   rapid movement of water that can pick up unprotected soil and carry it away

scrubber (SKRUHB uhr)   a device that removes dangerous particles from smoke as it passes up a smokestack

soil erosion (SOYL ee ROH zhuhn)   the loss of topsoil after roots and other supports are destroyed

solar energy (SOH luhr IHN uhr jee)   energy from the sun

stunt (STUHNT)   cause to grow abnormally small, crooked, or frail

sulfur dioxide (SUHL fuhr dy AHK syd)   a strong acid found in car exhaust

sulfuric acid (SUHL)   an acid found in the smoke of coal-burning power plants and factories

wetlands (WEHT landz)   marshes, swamps, bogs, ponds, and lakes